Alex Shneyer

Psychologist Drawing People

Alex Shneyer

Psychologist Drawing People

Napkin Sketches

Drugoe Reshenie

Impressum / Imprint
Bibliografische Information der Deutschen Nationalbibliothek: Die Deutsche Nationalbibliothek verzeichnet diese Publikation in der Deutschen Nationalbibliografie; detaillierte bibliografische Daten sind im Internet über http://dnb.d-nb.de abrufbar.

Bibliographic information published by the Deutsche Nationalbibliothek: The Deutsche Nationalbibliothek lists this publication in the Deutsche Nationalbibliografie; detailed bibliographic data are available in the Internet at http://dnb.d-nb.de.

Coverbild / Cover image: www.ingimage.com

Verlag / Publisher:
Drugoe Reshenie
ist ein Imprint der / is a trademark of
OmniScriptum Publishing Group
Contact: ICS Morebooks! Marketing SRL, 4, Industriala street, 3100 Balti, Republic of Moldova
Email: info@omniscriptum.com

Herstellung: siehe letzte Seite /
Printed at: see last page
ISBN: 978-3-330-33774-9

My dear Reader and Viewer!

Take this book with you.
Being on a trip, it is sometimes more pleasant to view pictures, leaf after leaf, than to read small font of text.
The present book has gathered various impressions, which are so immediate and thoughtless, that they have jumped on my papers, and stuck there. Among them you will find genre pictures that I caught simply in a cafeteria, on a beach, at a train station, in an airport — and then just carelessly lined out on a napkin. You will also find some thoughts, trying to catch up the impressions; these thoughts become a drawing after a while.
I know that if I let my hand move freely on a piece of paper, and then contemplate the result, I will see the chaotic dashes and lines as if by magic starting to form an image. Then I only have to erase all excess.
In fact, you see a dialogue with Unconscious in front of you.
And more — delight and need of a child — to confide...
May be you should try, too?

Sketches of the cities I like to visit play a special part in my works. It is a way I can express my gratitude for welcoming, warm and cosy atmosphere.

To be continued...

Yours faithfully,
artist, illustrator,
psychologist Oleksiy
Shneyer.
Or simply
-ША-

P.S.
If you want to become
acquainted with the author
and buy any of my works,
you can do it here:

http://alex-shneyer.blogspot.com

Napkin Sketches Series — the biggest and the most disordered one. Just sketches of life...

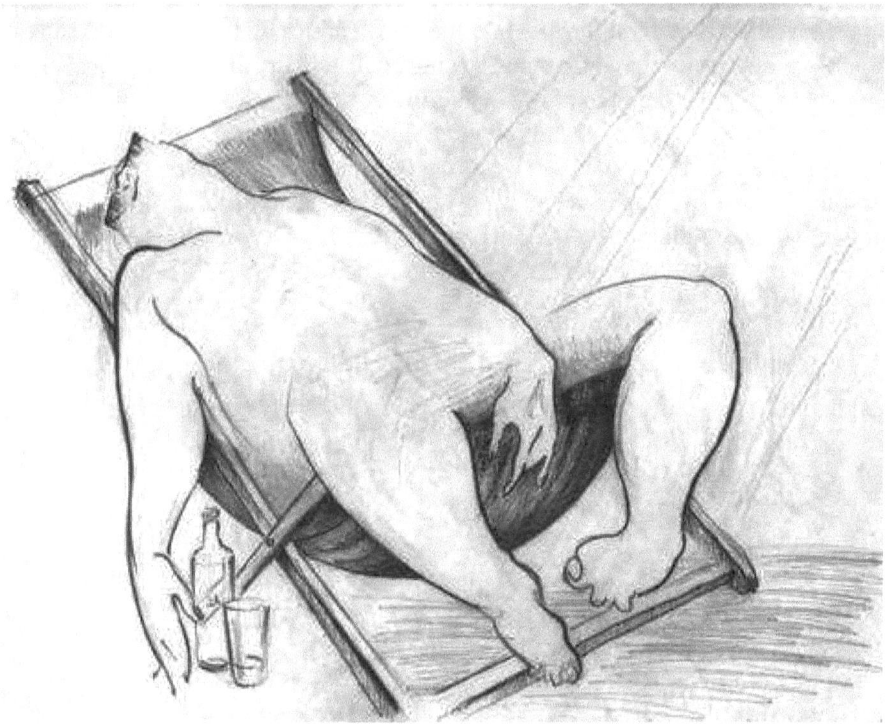

Chilling

Napkin Sketches

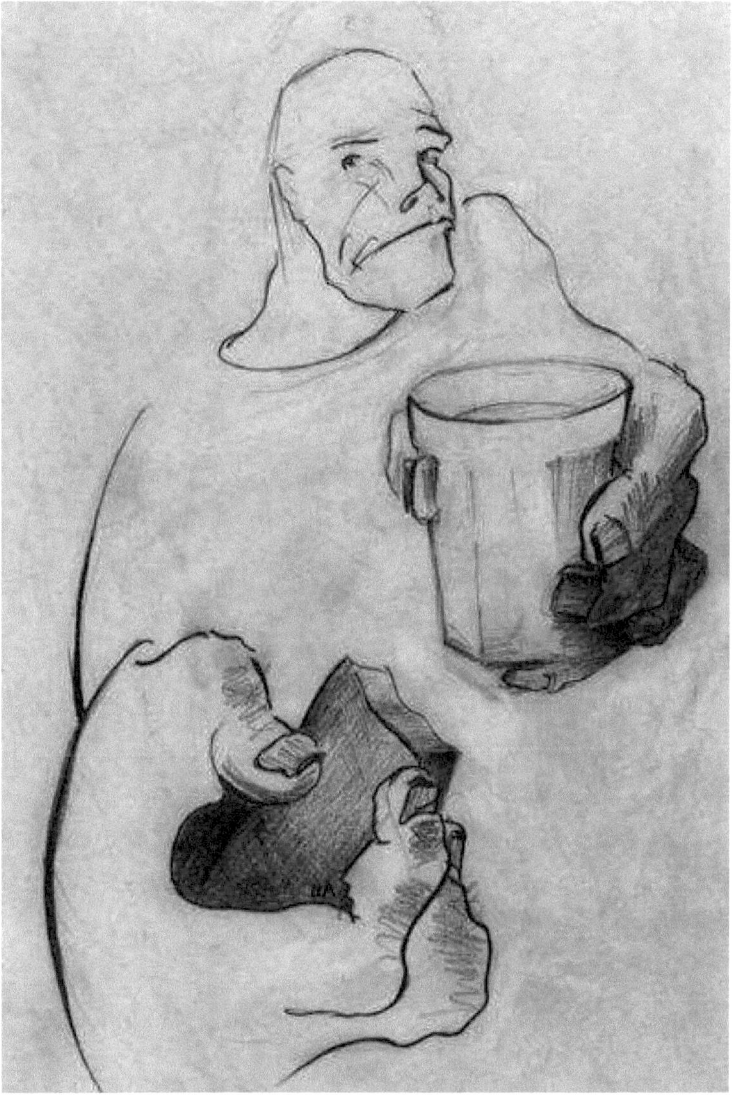

Will You Drink with Me?

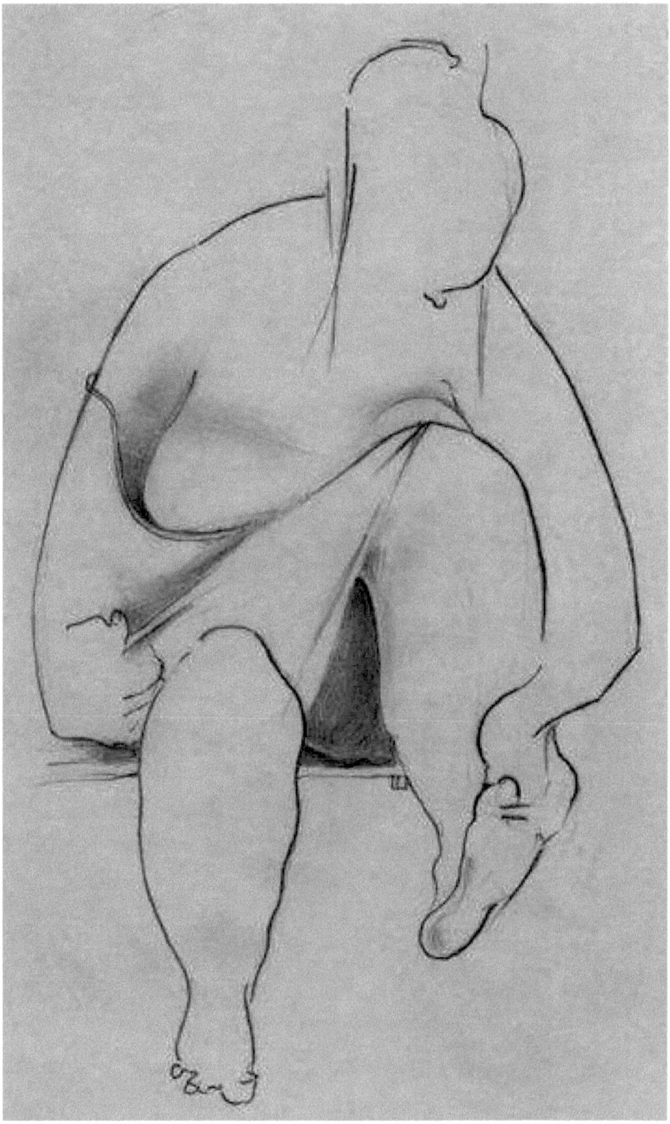

Gonna Have to Pedicure Again

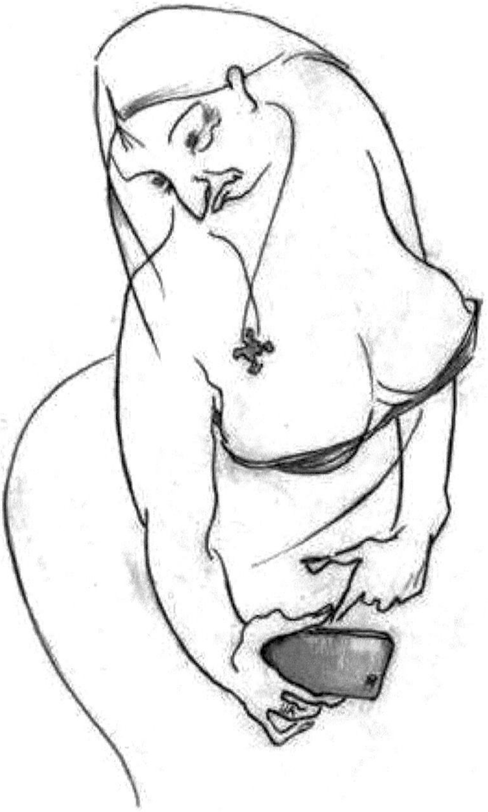

Good Angle For A Selfie

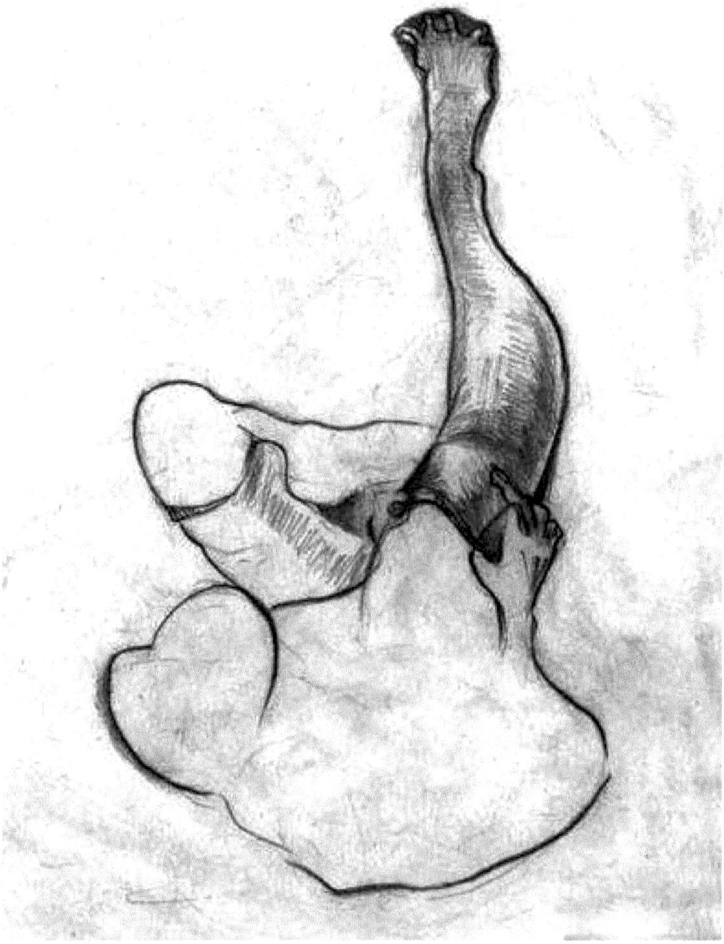

A Stocking

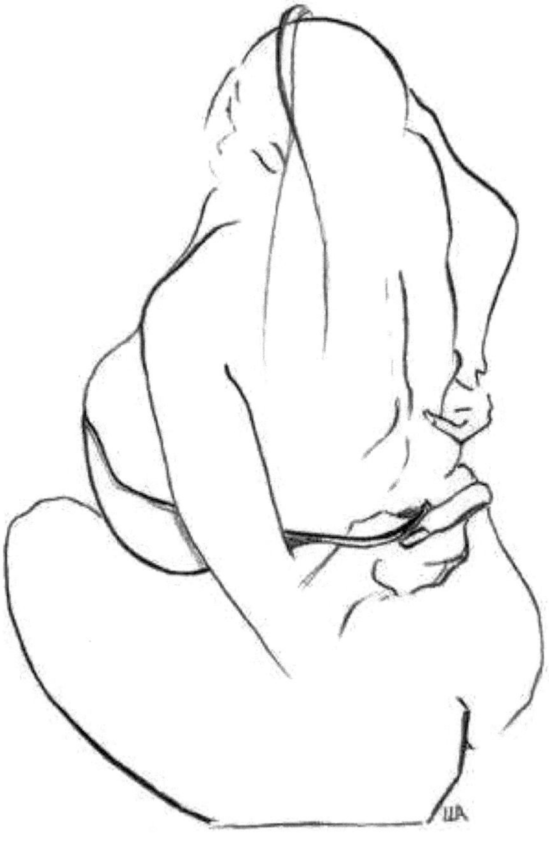

Topless

Ommmmm....

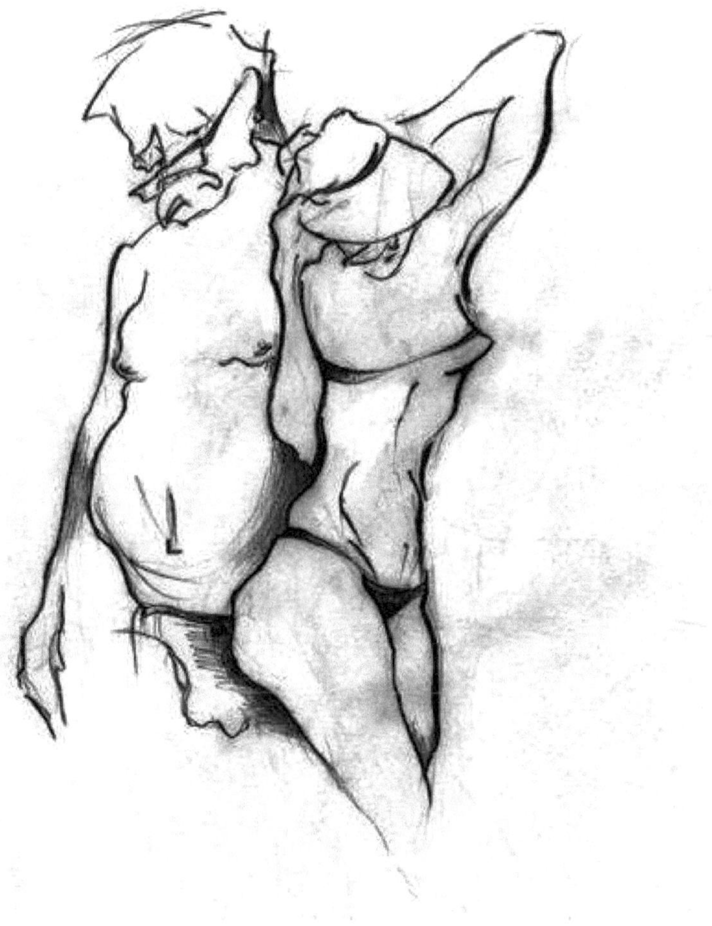

A Couple on the Beach

A Boring Holiday

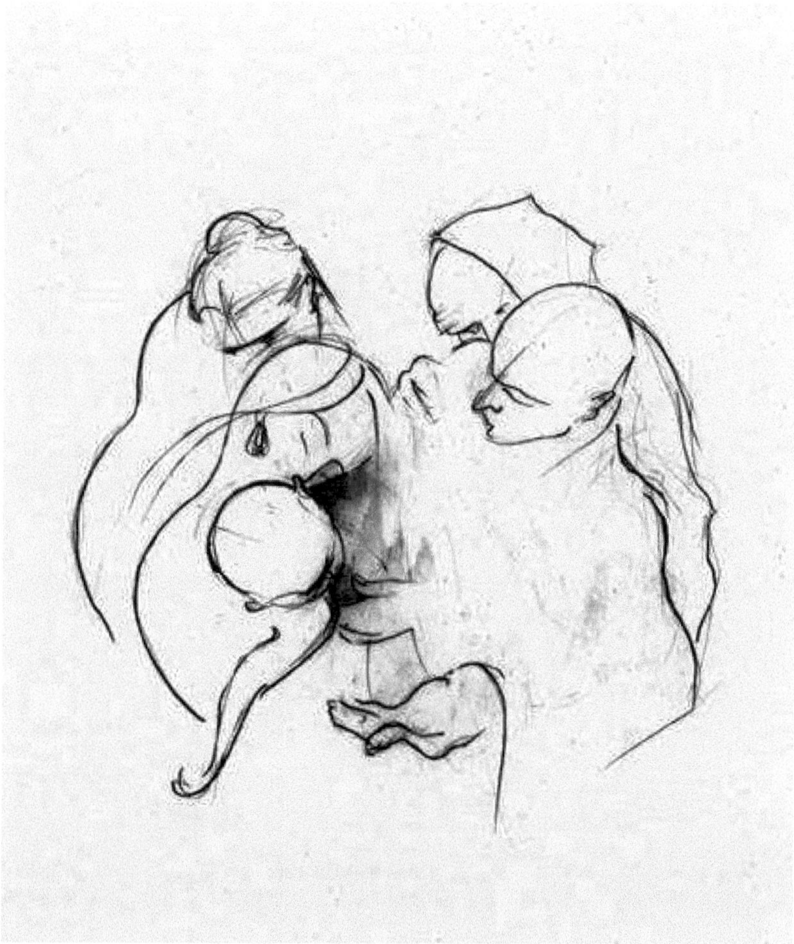

Getting to Know a New Dad

Man-to-Man Talk

A Deep Thinker

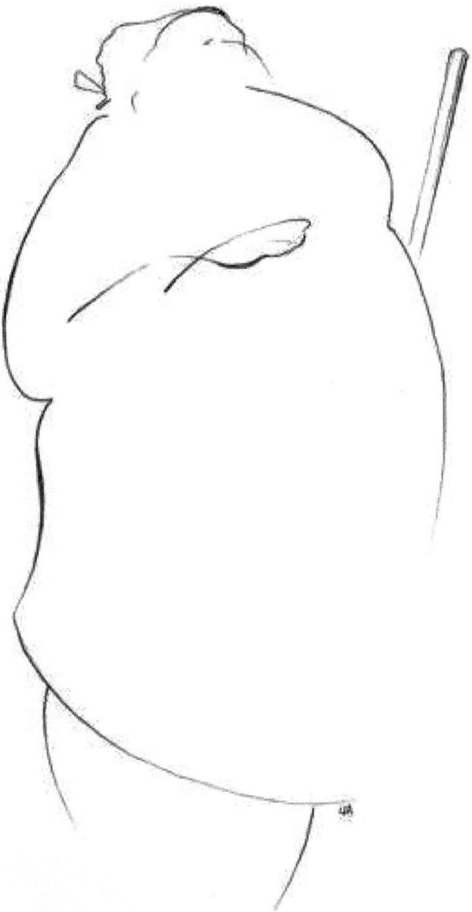

Nanny

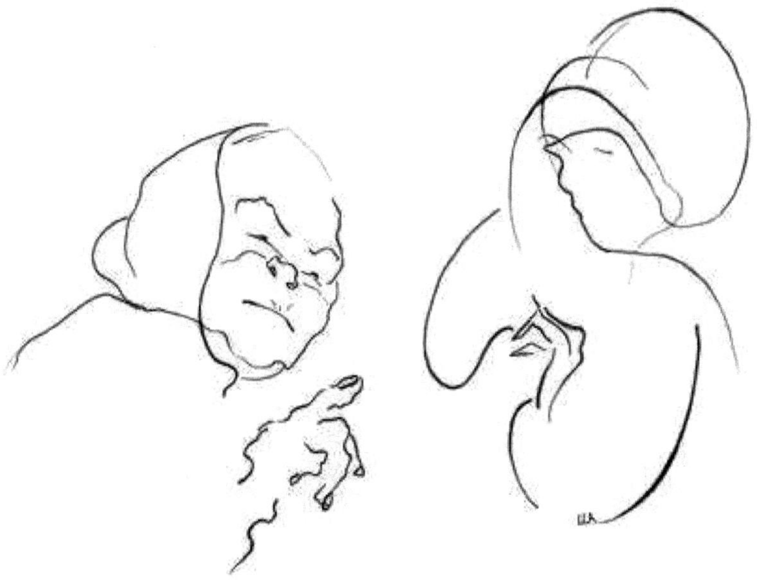

A Homily

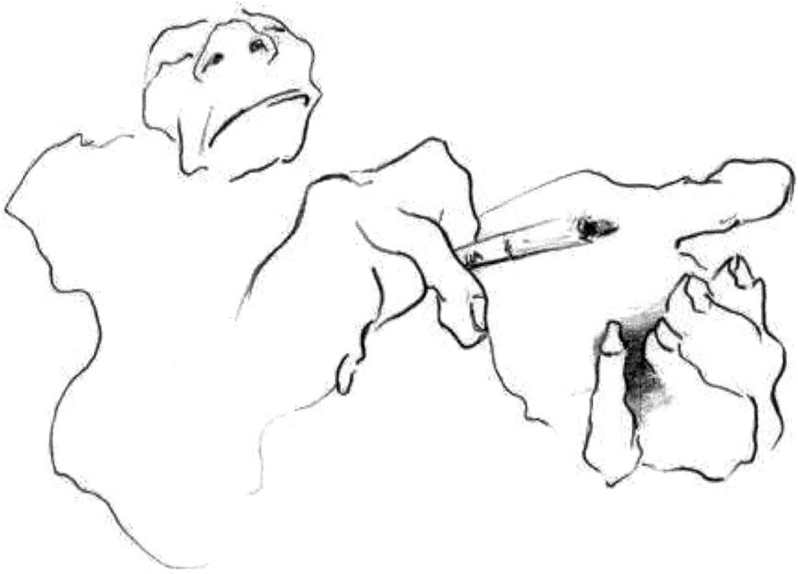

An Old Smoker

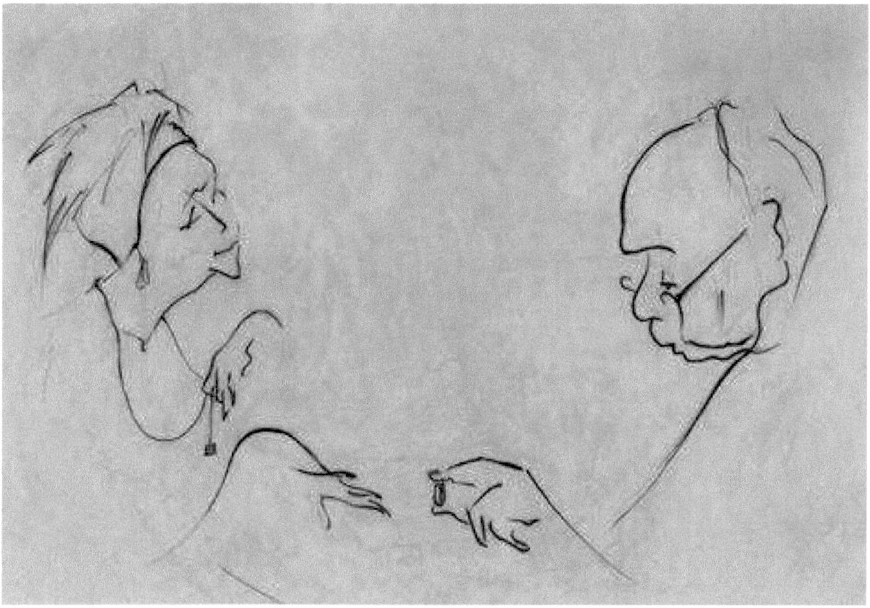

A Proposal

Tango

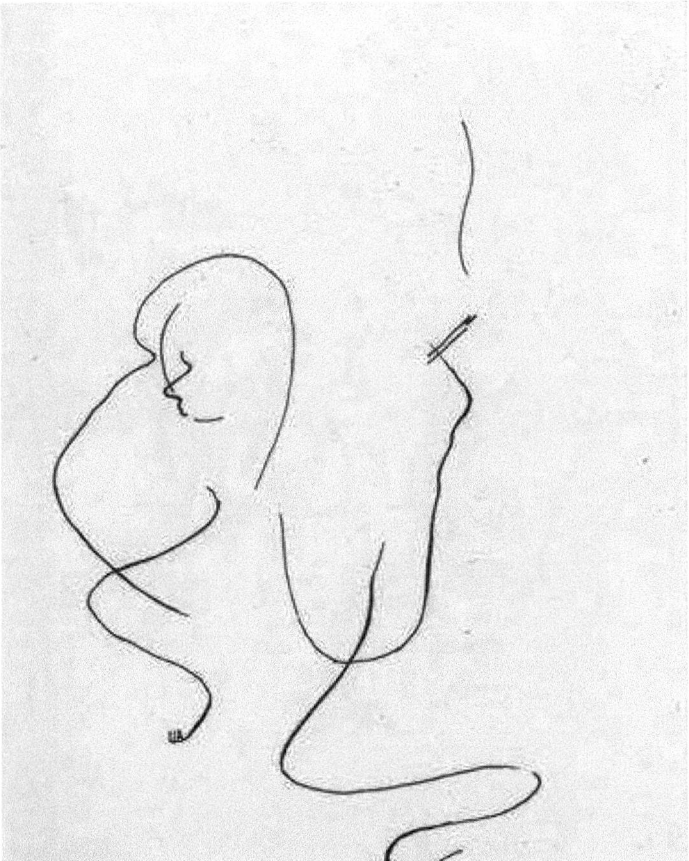

A Cigarette

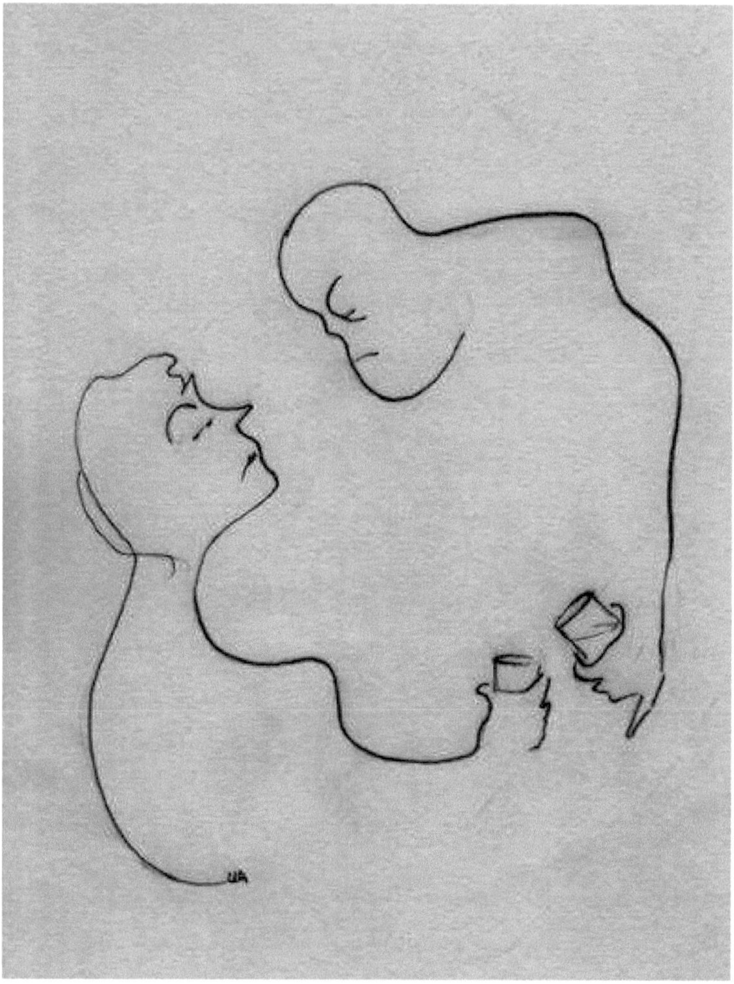

Let's drink to our
acquaintance

A Night Conversation

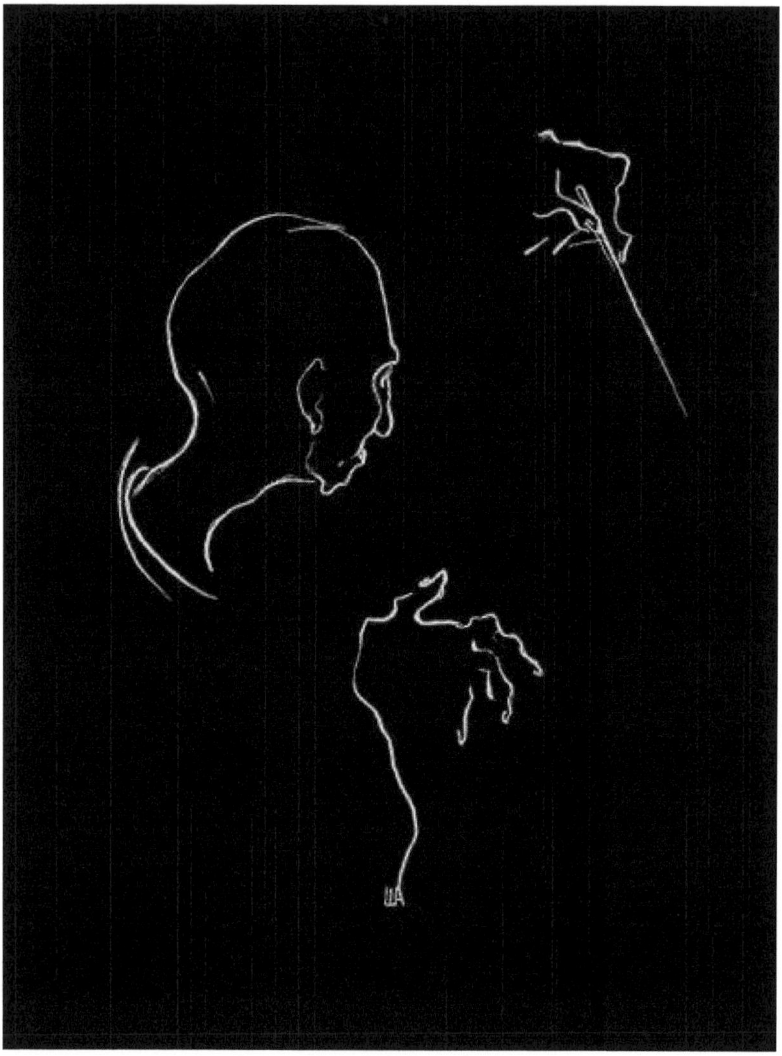

A Musical Director

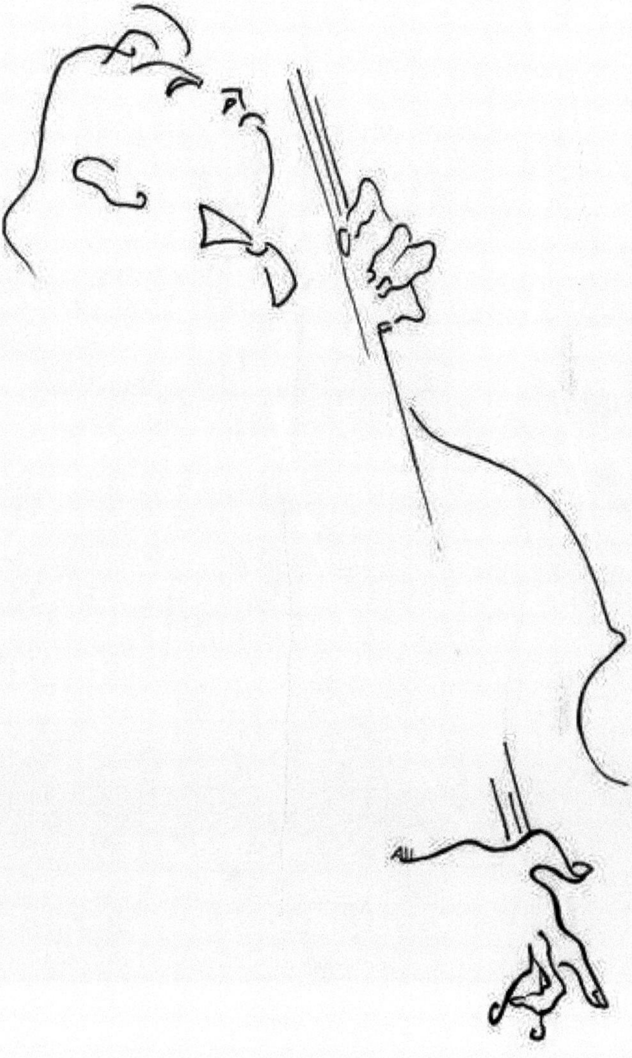

A Contrabass

A Flute

Musicians

A Flute

Musicians

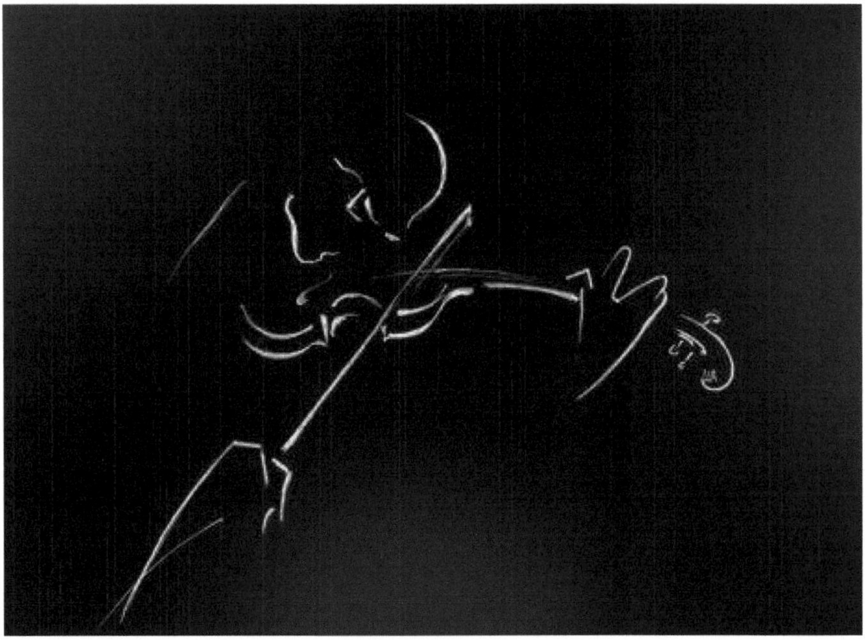

A Violinist

Musicians

Playing Jazz

A Fake Orchestra

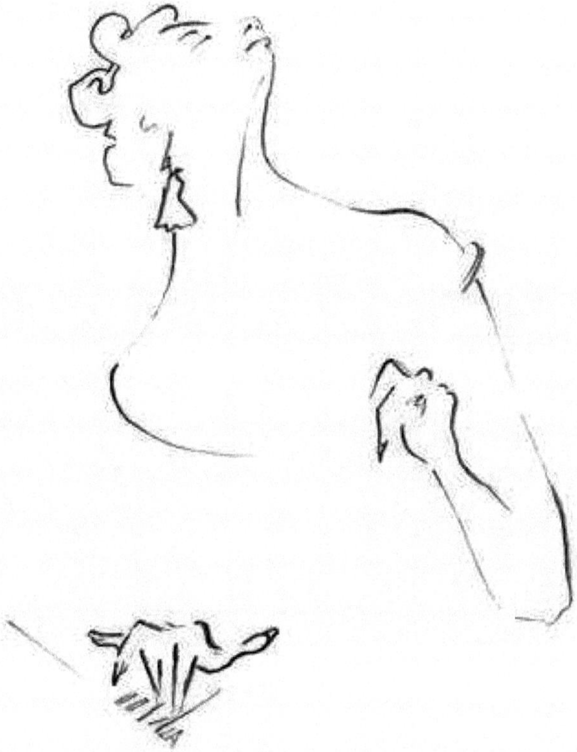

A Music Teacher

Musicians

A Random Melody

Cats Series

They have some magnetic charm...

Cats

Cats

Cats

Cats

Cats

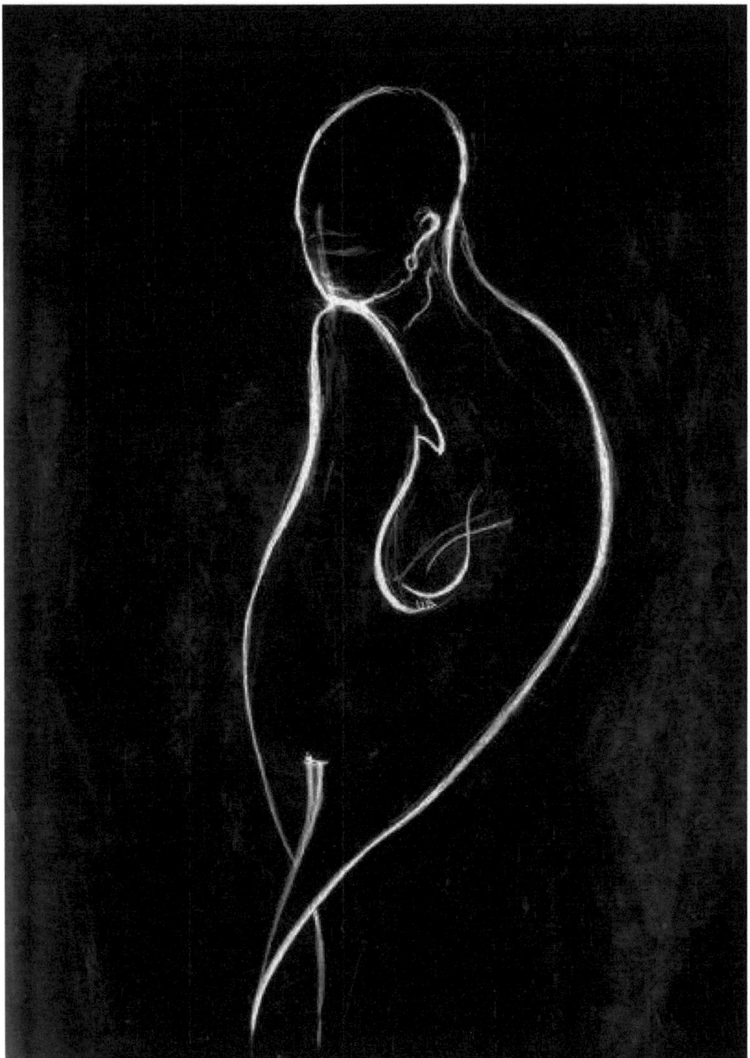

Cats

Party Series
Things like these happen with everybody...

A sudden compliment

Party

Party

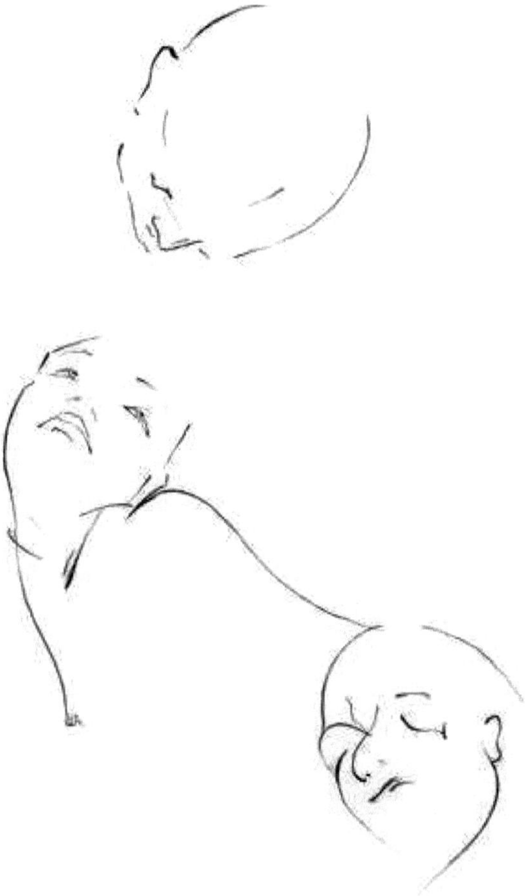

Last cocktail

Party

Party

Dance Series

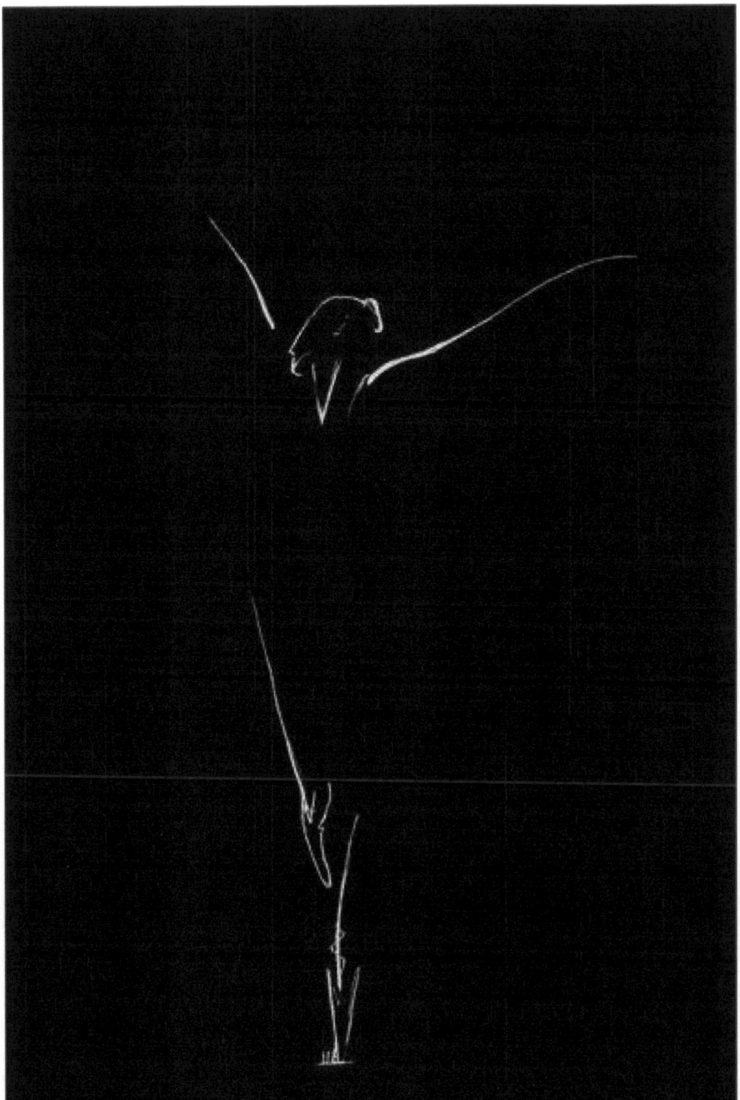

Dance

Dance

Dance

Dance

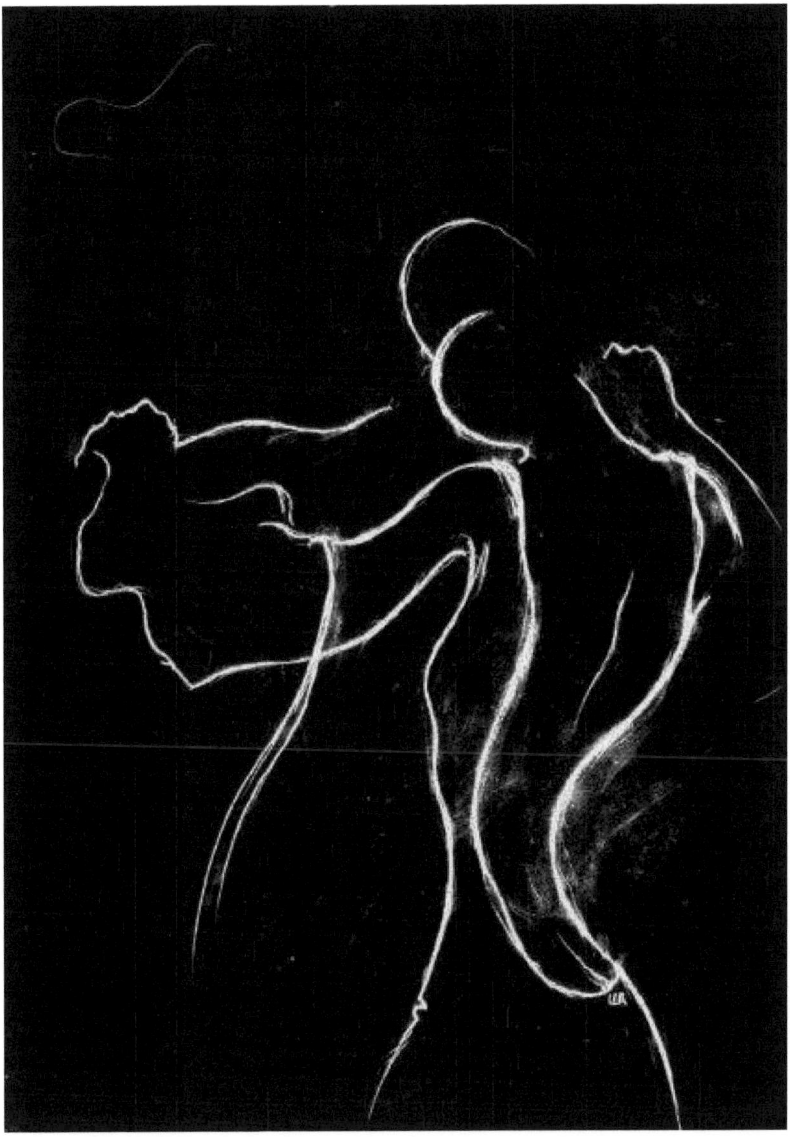

Passion Lines Series

Not for children and hypocrites!

Passion lines

Passion lines

Passion lines

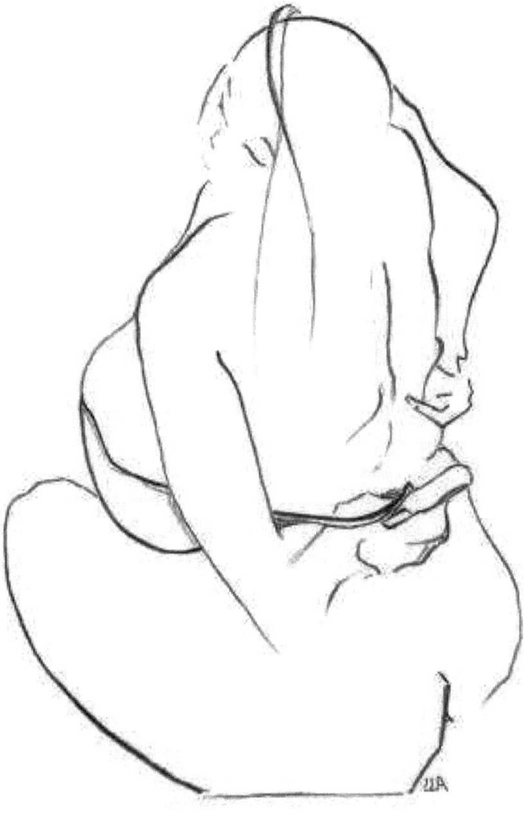

Passion lines

Passion lines

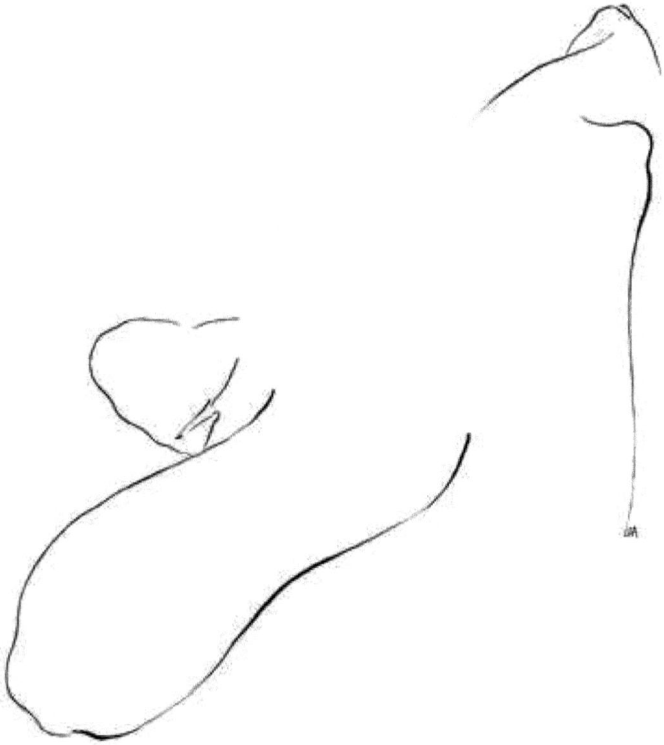

Passion lines

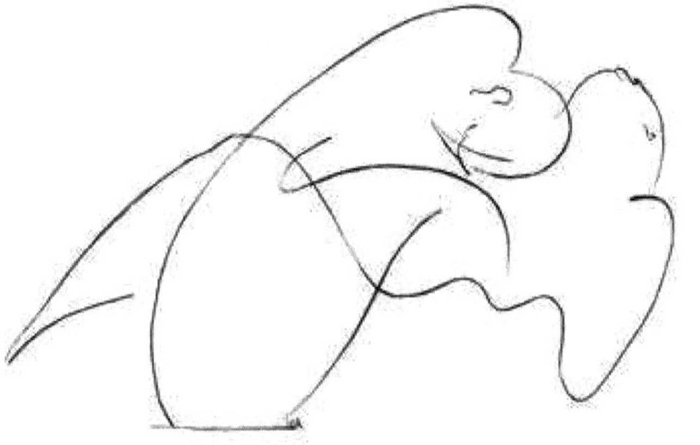

Passion lines

Passion lines

Passion lines

Passion lines

Morning Series

Morning

Morning

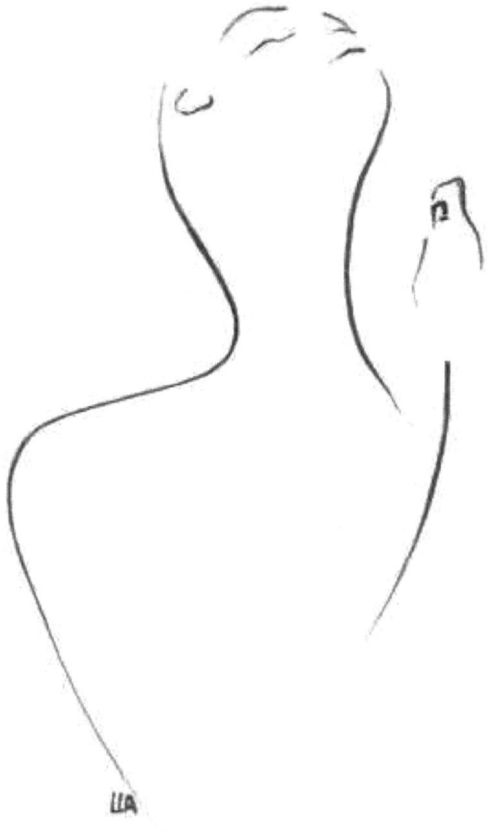

Morning

Morning

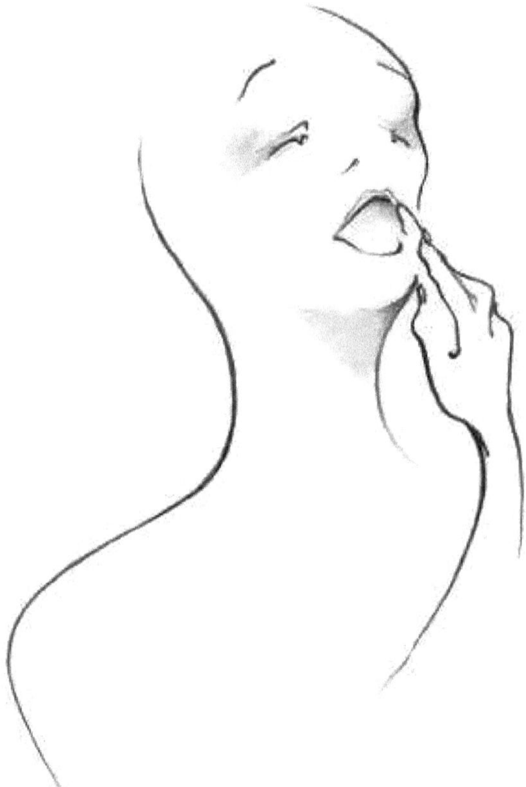

Morning

Morning

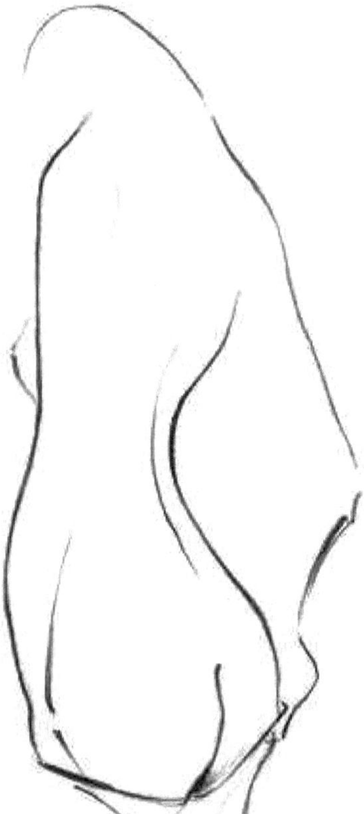

Morning

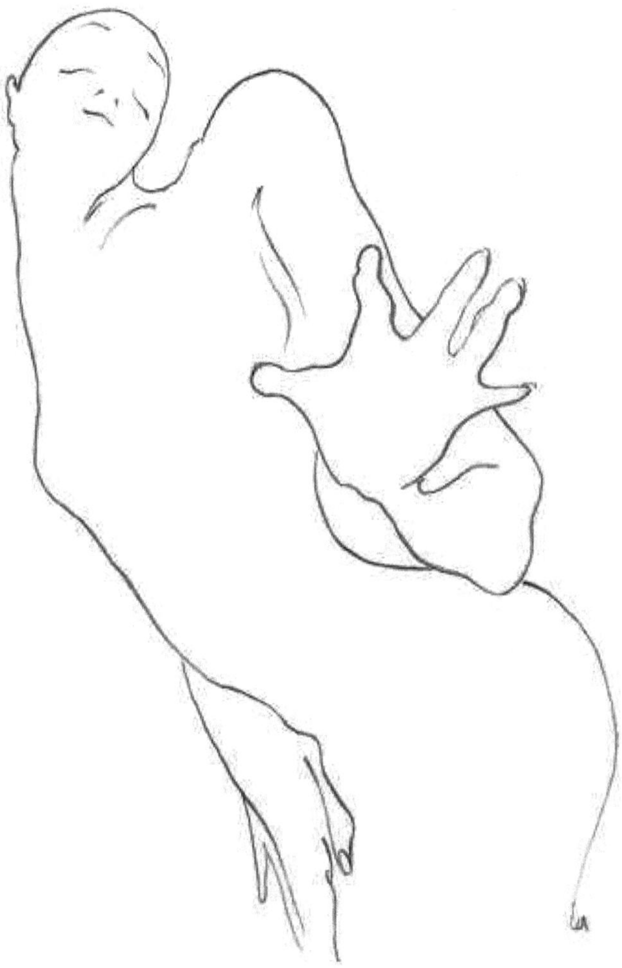

In a Jewish House Series

A Supper

A Self Portrait

In Memory of the Holocaust

Don't cry

An Old Tailor

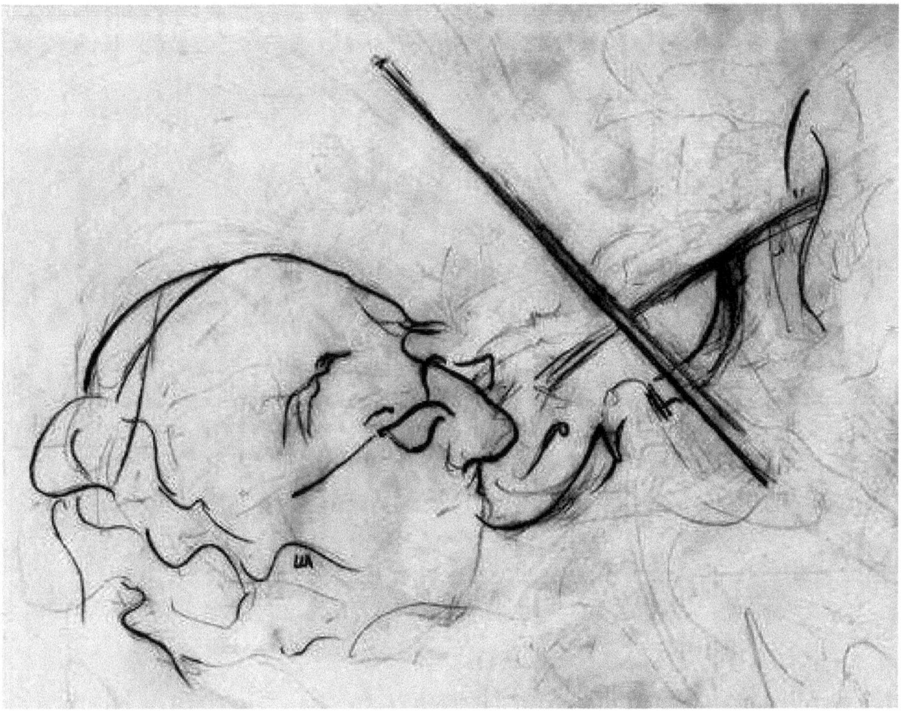

A Violinist

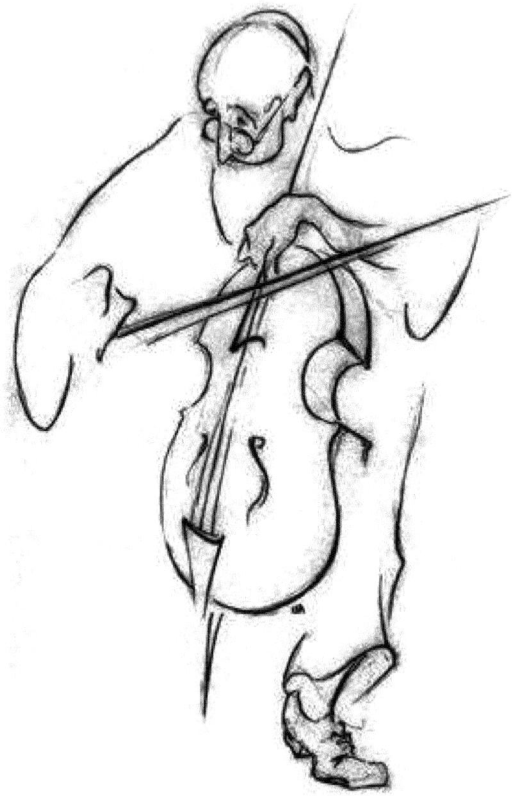

A Cello

Cityscapes Series
Kyiv

Cityscapes

Kyiv

Cityscapes

Kyiv

Cityscapes

Kyiv

Cityscapes

Kyiv

Cityscapes

Lviv

Cityscapes

Lviv

Cityscapes

Tbilisi

Cityscapes

Tbilisi

Cityscapes

Tbilisi

Cityscapes

Tbilisi

Cityscapes

Tbilisi

Cityscapes

Tbilisi

About the Author

Oleksiy Shneyer

Was born in 1956 in Kyiv, USSR and immediately started to draw. But for a while it was only a hobby.

Graduated from Polytechnic Institute of Novocherkassk (Novocherkassk, USSR). Before the Soviet breakup worked as design engineer and electronics engineer.

After the breakup resumed studying:

- Baltic Pedagogical Academy (Saint-Petersburg, Russia, in practical psychology)
- Psychotherapy School of Kyiv (Kyiv, Ukraine, in psychoanalysis)
- Suicidology School of Kyiv (Kyiv, Ukraine). Worked in his specialization for a long time: consulting and leading analytic groups.

The first exhibition of the author's artwork took place in Kyiv on July 2, 2014.

Today Oleksiy Shneyer cooperates with leading Ukranian publishing house SUMMIT-BOOK as an illustrator, sells his artwork,

keeps a blog

http://alex-shneyer.blogspot.com

and a Facebook page:

https://www.facebook.com/altapsy

Always open for dialog and cooperation

Index

MIX
Papier aus verantwortungsvollen Quellen
Paper from responsible sources
FSC® C105338

Printed by Books on Demand GmbH, Norderstedt / Germany